# Farm the Future Optimizing Agriculture with IoT Technologies

## Myles Grace

**Copyright © [2023]**

## Title: Farm the Future Optimizing Agriculture with IoT Technologies

## Author's: Myles Grace



This book was printed and published by [Publisher's: **Myles Grace**] in [2023]

**ISBN:**

# TABLE OF CONTENT

# Chapter 1: Introduction to Smart Agriculture and IoT Technologies

## The Importance of Agriculture in the Modern World

In today's rapidly advancing world, the importance of agriculture cannot be overstated. With the global population expected to reach 9.7 billion by 2050, the need to feed a growing population becomes more pressing than ever. Agriculture plays a vital role in ensuring food security, economic growth, and environmental sustainability. This subchapter explores the significance of agriculture in the modern world and how the integration of Internet of Things (IoT) technologies can optimize its practices.

Food Security: Agriculture is the backbone of food production, providing nourishment to billions of people worldwide. With the increasing demand for food, it is crucial to embrace innovative technologies that enhance production efficiency and yield. IoT technologies enable farmers to monitor weather patterns, soil conditions, and crop health in real-time, allowing for precise and timely decision-making. By optimizing irrigation, fertilization, and pest control, farmers can maximize crop yield and ensure a steady food supply for the growing population.

Economic Growth: The agricultural sector is a significant contributor to the global economy. It provides employment opportunities for millions of people and supports rural development. By incorporating IoT technologies, farmers can streamline their operations, reduce waste, and improve productivity. Automated systems, such as smart irrigation and precision farming, optimize resource allocation, resulting in higher yields and increased profitability. Furthermore,

IoT-enabled supply chain management ensures efficient distribution, reducing food waste and improving market access for farmers.

Environmental Sustainability: With climate change posing a threat to agricultural practices, sustainability becomes paramount. IoT technologies offer solutions to mitigate environmental impacts and promote sustainable farming practices. By continuously monitoring soil moisture, temperature, and nutrient levels, farmers can minimize water and fertilizer usage, reducing the risk of pollution and conserving natural resources. Additionally, IoT sensors can detect early signs of pest infestations, allowing farmers to take prompt action, minimizing the need for harmful pesticides.

In conclusion, agriculture remains a critical sector in the modern world, addressing the challenges of food security, economic growth, and environmental sustainability. The integration of IoT technologies into agricultural practices enables farmers to optimize production, reduce waste, and ensure sustainable farming practices. By embracing these innovations, we can work towards a future where agriculture meets the demands of a growing population while preserving our planet's resources. Whether you are a farmer, consumer, or simply someone interested in the Internet of Things, understanding the importance of agriculture and its potential with IoT technologies is crucial for a sustainable future.

## Overview of IoT Technologies and Their Applications in Agriculture

The rapid advancement of technology has revolutionized various industries, and agriculture is no exception. In recent years, the Internet of Things (IoT) has emerged as a game-changing technology with immense potential to optimize agricultural practices. This subchapter aims to provide an overview of IoT technologies and their applications in agriculture, offering valuable insights to a diverse audience.

IoT technologies refer to the network of interconnected devices, sensors, and software that enable the collection, analysis, and exchange of data. In agriculture, these technologies have proven to be invaluable in enhancing efficiency, productivity, and sustainability. By integrating IoT solutions into farming practices, farmers can make more informed decisions, reduce waste, and maximize yields.

One of the most prominent applications of IoT in agriculture is precision farming. With the help of sensors and data analytics, farmers can monitor and control various factors such as soil moisture, temperature, and nutrient levels in real-time. This allows for precise irrigation, fertilization, and pest control, resulting in optimized resource utilization and improved crop health.

Livestock management is another area where IoT technologies have made significant contributions. Smart sensors can monitor the health, behavior, and location of animals, providing valuable data for disease detection, early warning systems, and optimized feeding schedules. Additionally, automated systems can regulate temperature, ventilation, and lighting in animal housing, creating a comfortable and stress-free environment.

IoT technologies have also facilitated the development of smart greenhouses and vertical farming systems. By integrating sensors, actuators, and control systems, these innovative setups can maintain optimal growing conditions for crops, independent of external factors. This enables year-round production, minimizes water and energy consumption, and reduces the environmental footprint of agriculture.

In addition to farm-level applications, IoT technologies have also revolutionized supply chain management in agriculture. By tracking and monitoring produce from farm to market, stakeholders can ensure quality control, minimize waste, and optimize logistics. Real-time data on temperature, humidity, and storage conditions enable timely interventions, reducing post-harvest losses and enhancing food safety.

Overall, the potential of IoT technologies in agriculture is vast and transformative. By harnessing the power of interconnected devices and data analytics, farmers can make agriculture more sustainable, efficient, and resilient. This subchapter provides a comprehensive overview of these technologies and their applications, appealing to a broad audience interested in the Internet of Things and its potential in the agricultural sector.

# Chapter 2: IoT Solutions for Soil Monitoring and Management

## Understanding Soil Health and Nutrient Levels

Soil health and nutrient levels are vital components of successful agriculture. In this subchapter, we will explore the importance of soil health, the role of nutrients in plant growth, and how Internet of Things (IoT) technologies can optimize agriculture by monitoring and improving soil conditions.

Soil health is the foundation of any agricultural system. It refers to the overall well-being of the soil, including its physical, chemical, and biological properties. Healthy soil is rich in nutrients, has good structure and water-holding capacity, and supports a diverse community of beneficial organisms. By understanding and improving soil health, farmers can enhance crop productivity, reduce the need for synthetic inputs, and promote environmental sustainability.

Nutrients play a crucial role in plant growth and development. Plants require a range of essential nutrients, including macronutrients like nitrogen, phosphorus, and potassium, as well as micronutrients like iron, zinc, and manganese. Imbalances or deficiencies in these nutrients can lead to reduced crop yields, poor plant health, and susceptibility to pests and diseases. Regular monitoring of nutrient levels in the soil is therefore necessary to ensure optimal plant nutrition and productivity.

IoT technologies offer exciting opportunities for farmers to monitor and manage soil health and nutrient levels more effectively. These technologies involve the use of sensors, data analytics, and cloud computing to collect and analyze real-time data on soil conditions. Soil moisture sensors can measure the water content in the soil, helping farmers optimize irrigation practices and conserve water. Nutrient sensors can provide accurate and timely information about the nutrient levels in the soil, enabling farmers to apply fertilizers precisely where and when they are needed, minimizing waste and environmental impact.

Additionally, IoT technologies can track and analyze data on soil temperature, pH levels, and organic matter content, providing valuable insights into soil health. By integrating these data with weather forecasts and crop growth models, farmers can make informed decisions about planting, irrigation, and nutrient management, optimizing resource use and maximizing yields.

In conclusion, understanding soil health and managing nutrient levels are essential for sustainable and productive agriculture. IoT technologies offer innovative solutions to monitor and optimize soil conditions, empowering farmers to make informed decisions and improve their farming practices. By harnessing the power of IoT, we can pave the way for a more efficient and environmentally friendly farm of the future.

**IoT Sensors for Soil Moisture Monitoring**

In today's technologically advanced world, the Internet of Things (IoT) has revolutionized various industries, and agriculture is no exception. The integration of IoT technologies has transformed traditional farming practices into smart and efficient systems. One such application is the use of IoT sensors for soil moisture monitoring, which has proven to be a game-changer in optimizing agriculture.

Soil moisture plays a crucial role in determining the health and productivity of crops. Insufficient or excessive moisture can lead to crop failure, resulting in significant financial losses for farmers. Therefore, monitoring soil moisture levels accurately is essential to make informed decisions regarding irrigation and crop management.

IoT sensors provide a cost-effective and efficient solution for real-time soil moisture monitoring. These sensors are embedded in the ground at various depths, collecting data on moisture levels. The data is then transmitted wirelessly to a central system, which analyzes and presents the information in a user-friendly format. Farmers can access this data from anywhere, using smartphones or computers, making it convenient and accessible.

The benefits of using IoT sensors for soil moisture monitoring are numerous. Firstly, it enables farmers to optimize water usage by providing real-time information on soil moisture levels. This helps prevent over- or under-irrigation, ensuring that crops receive the necessary amount of water, thus conserving water resources and reducing costs.

Secondly, IoT sensors allow farmers to detect and address potential issues promptly. For example, if the soil moisture levels drop below a certain threshold, an alert can be sent to the farmer, indicating the need for irrigation. This proactive approach helps prevent crop damage and ensures timely intervention.

Furthermore, IoT sensors enable farmers to make data-driven decisions. By analyzing historical data on soil moisture levels, farmers can identify patterns and trends, allowing them to fine-tune irrigation schedules and optimize crop yields. This data can also be combined with other environmental factors, such as weather forecasts, to create comprehensive models for informed decision-making.

In conclusion, the integration of IoT sensors for soil moisture monitoring is transforming agriculture by providing farmers with real-time and accurate data. This technology empowers farmers to optimize water usage, prevent crop damage, and make data-driven decisions for improved crop productivity. With IoT sensors, the farm of the future is becoming a reality, where technology and agriculture converge to create sustainable and efficient farming practices.

## Smart Fertilizer Management Systems

In today's world where technology is rapidly advancing, the agriculture industry is also benefitting from the integration of Internet of Things (IoT) technologies. One such innovation is the development of Smart Fertilizer Management Systems, which revolutionizes the way farmers monitor and optimize their fertilizer usage. These systems utilize IoT sensors, data analytics, and automation to enhance crop productivity while minimizing environmental impact.

Traditionally, farmers applied fertilizers based on guesswork, past experience, or general recommendations. However, this approach often led to over or under-fertilization, resulting in wastage of resources and negative consequences for the environment. Smart Fertilizer Management Systems address this issue by providing real-time data and insights on soil conditions, crop needs, and fertilization requirements.

With the help of IoT sensors, these systems continuously monitor soil moisture levels, temperature, pH levels, and nutrient content. The data collected is then transmitted to a central platform where it is analyzed using advanced algorithms. By combining this information with weather forecasts, crop growth models, and historical data, farmers can make informed decisions about when and how much fertilizer to apply.

By leveraging automation capabilities, Smart Fertilizer Management Systems can precisely control the amount and timing of fertilizer application. This eliminates the risk of human error and ensures that crops receive the optimal nutrients at the right time. Additionally, these systems can be integrated with irrigation systems, allowing for efficient water management and minimizing water wastage.

The benefits of adopting Smart Fertilizer Management Systems are numerous. Firstly, farmers can achieve higher crop yields and improved quality, leading to increased profitability. Secondly, these systems reduce environmental impact by minimizing fertilizer runoff and groundwater contamination. By applying the right amount of fertilizer, farmers can also save costs by reducing fertilizer purchases and minimizing unnecessary applications.

Moreover, Smart Fertilizer Management Systems enable farmers to comply with regulatory requirements and sustainability standards. By monitoring and documenting their fertilizer usage, farmers can provide transparent information to stakeholders and consumers, building trust and reinforcing their commitment to sustainable practices.

In conclusion, Smart Fertilizer Management Systems represent a significant advancement in agriculture, enabled by IoT technologies. These systems empower farmers with real-time data, data analytics, and automation capabilities to optimize their fertilizer usage, increase crop productivity, and minimize environmental impact. By adopting these systems, farmers can embrace sustainable practices, improve profitability, and contribute to the future of agriculture.

## Optimizing Irrigation Practices with IoT

Irrigation plays a vital role in agriculture by providing water to crops to ensure healthy growth and maximum yield. However, traditional irrigation practices often involve inefficiencies and waste, leading to unnecessary water consumption and increased costs for farmers. In recent years, the advent of Internet of Things (IoT) technologies has revolutionized the way we approach irrigation, offering innovative solutions to optimize water usage and enhance agricultural productivity.

IoT devices, such as sensors and actuators, can be seamlessly integrated into irrigation systems to monitor and control water flow in real-time. These devices collect data on various parameters, including soil moisture levels, weather conditions, and crop water requirements, and transmit it to a central hub or cloud-based platform. This data is then analyzed using advanced algorithms and machine learning techniques to provide actionable insights and automate irrigation processes.

By leveraging IoT technologies, farmers can achieve precise irrigation management tailored to the specific needs of their crops. Real-time data on soil moisture levels helps farmers determine the optimal time and amount of water required, preventing under or over-irrigation. This not only conserves water but also ensures that crops receive the necessary nutrients for healthy growth. Additionally, IoT-enabled irrigation systems can adjust water flow based on weather forecasts, preventing unnecessary irrigation during periods of rainfall.

The benefits of optimizing irrigation practices with IoT extend beyond water conservation. Farmers can also reduce energy consumption by using IoT devices to automate irrigation processes, eliminating the need for manual intervention. Moreover, IoT platforms provide comprehensive analytics and reporting, allowing farmers to track water usage, identify areas of improvement, and make data-driven decisions to enhance overall efficiency.

Furthermore, IoT technologies enable remote monitoring and control of irrigation systems, empowering farmers to manage their operations from anywhere at any time. This level of flexibility and convenience enhances productivity and reduces the labor-intensive nature of traditional irrigation practices.

In conclusion, optimizing irrigation practices with IoT holds tremendous potential for revolutionizing agriculture. By harnessing the power of IoT devices and advanced analytics, farmers can reduce water consumption, conserve energy, improve crop yields, and ultimately contribute to sustainable and efficient farming practices. The future of agriculture lies in embracing IoT technologies to optimize irrigation practices and ensure a prosperous and greener future for all.

# Chapter 3: Enhancing Crop Growth and Yield with IoT Technologies

## Monitoring Plant Health and Growth Parameters

In today's rapidly evolving world, the integration of Internet of Things (IoT) technologies has revolutionized various industries, and agriculture is no exception. The Farm of the Future: Optimizing Agriculture with IoT Technologies is a comprehensive guide that introduces the concept of using IoT to enhance farming practices, increase productivity, and ensure sustainable agriculture. This subchapter, "Monitoring Plant Health and Growth Parameters," delves into the crucial role IoT plays in monitoring and managing the health and growth of plants, making it a valuable read for everyone interested in the Internet of Things.

The subchapter begins by highlighting the importance of plant health and growth parameters in ensuring successful crop production. It emphasizes the significance of early detection and prevention of diseases, pests, nutrient deficiencies, and environmental stressors that can significantly impact plant health. By utilizing IoT technologies, farmers and agricultural experts can now remotely monitor and analyze vital parameters such as temperature, humidity, soil moisture, light intensity, and nutrient levels in real-time.

The content then explores the various IoT sensors and devices that are used to collect data on plant health and growth parameters. From soil moisture sensors to aerial drones equipped with multispectral cameras, these technologies provide accurate and detailed information about the condition of crops. The chapter explains how these devices are seamlessly integrated into existing farming practices and how the collected data is transmitted to a central system for analysis.

Furthermore, the subchapter discusses the significance of data analytics and artificial intelligence in interpreting the collected data. Advanced algorithms can process vast amounts of information and provide valuable insights, helping farmers make informed decisions about irrigation, fertilization, and pest control. By utilizing IoT technologies, farmers can optimize their resource usage, reduce waste, and improve overall crop yield.

The content also acknowledges the potential challenges and limitations of implementing IoT in agriculture. It highlights the importance of data security and privacy, as well as the need for proper training and support for farmers to effectively use these technologies.

Overall, this subchapter serves as an essential resource for everyone interested in IoT technologies and their application in agriculture. By monitoring plant health and growth parameters through IoT, farmers can significantly enhance their farming practices, increase productivity, and contribute to sustainable agriculture practices. The Farm of the Future: Optimizing Agriculture with IoT Technologies is a must-read for anyone seeking to understand the transformative potential of IoT in the agriculture industry.

## Precision Farming Techniques with IoT

Precision farming, also known as precision agriculture, is a modern farming approach that utilizes advanced technologies to optimize crop production and increase overall efficiency. In recent years, the integration of Internet of Things (IoT) technologies has revolutionized precision farming, allowing farmers to monitor and manage their farms more effectively than ever before.

IoT in precision farming involves the use of sensors, drones, and other smart devices that collect data on various environmental factors, such as soil moisture, temperature, and humidity. This data is then transmitted to a central system, which farmers can access from anywhere using their smartphones or tablets. By analyzing this data, farmers can make informed decisions about irrigation, fertilization, and pest control, thus minimizing waste and maximizing yields.

One of the key advantages of precision farming techniques with IoT is the ability to monitor crops in real-time. With the help of IoT devices, farmers can remotely monitor the health and growth of their crops, detecting any potential issues or anomalies. For instance, if there is a sudden drop in soil moisture levels, the IoT system can send an alert to the farmer, allowing them to take immediate action and prevent crop damage.

Furthermore, IoT technologies enable farmers to implement site-specific management practices. By collecting data on soil conditions, farmers can create detailed maps of their fields, highlighting areas that require specific treatments or interventions. This allows for targeted application of fertilizers and pesticides, reducing overall input costs and minimizing the environmental impact.

In addition to crop monitoring, IoT devices also play a crucial role in livestock management. Sensors attached to animals can track their health, behavior, and location, providing valuable insights to farmers. For example, if a cow's body temperature rises above normal, the IoT system can alert the farmer, signaling the potential onset of an illness. This early warning system allows for timely intervention, ensuring the well-being of the animals and preventing the spread of diseases.

Overall, precision farming techniques with IoT have the potential to transform the agricultural industry. By harnessing the power of IoT technologies, farmers can optimize resource allocation, improve productivity, and reduce environmental impact. Whether you are a farmer, a technology enthusiast, or simply curious about the future of agriculture, understanding the potential of IoT in precision farming is essential. Embracing these innovative solutions can lead to a more sustainable and efficient farming industry, benefiting not only farmers but also society as a whole.

## Remote Crop Monitoring and Management

In today's ever-evolving world, the Internet of Things (IoT) has revolutionized various industries, and agriculture is no exception. With the rise of IoT technologies, farmers and agronomists can now optimize their crop monitoring and management practices like never before. This subchapter will explore the significance of remote crop monitoring and management in the context of the Farm of the Future.

Remote crop monitoring involves the use of IoT devices, such as sensors and drones, to gather real-time data on crop conditions, soil moisture levels, temperature, humidity, and other crucial parameters. By collecting this data, farmers can gain valuable insights into the health and growth of their crops, enabling them to make informed decisions and take timely actions to ensure optimal yields.

One of the key advantages of remote crop monitoring is its ability to provide farmers with a comprehensive view of their fields, regardless of their location. Through IoT-enabled platforms, farmers can access data from multiple sensors spread across their fields, allowing them to monitor and manage their crops remotely. This technology eliminates the need for physical presence and reduces the time and effort required for manual inspections.

Furthermore, IoT technologies enable the automation of various crop management tasks. Smart irrigation systems, for instance, can be programmed to water crops based on real-time data from soil moisture sensors. This ensures that crops receive the right amount of water at the right time, reducing water wastage and optimizing water usage efficiency. Similarly, automated pest and disease monitoring systems can detect early signs of infestations, allowing farmers to take immediate action and prevent significant crop damage.

The benefits of remote crop monitoring and management extend beyond individual farms. By harnessing the power of IoT data, agronomists and researchers can analyze large-scale trends and patterns, leading to the development of more effective and sustainable agricultural practices. This collective knowledge can be shared with farmers worldwide, fostering collaboration and innovation in the agricultural sector.

In conclusion, remote crop monitoring and management using IoT technologies have the potential to revolutionize agriculture. By providing real-time data and insights, this technology empowers farmers to optimize their crop yields, conserve resources, and make informed decisions. The Farm of the Future embraces IoT and its benefits, creating a sustainable and efficient agricultural ecosystem for the benefit of everyone.

## Predictive Analytics for Crop Yield Optimization

In today's rapidly changing world, the agriculture industry is constantly seeking innovative ways to optimize crop yield and increase productivity. One such groundbreaking solution is the use of predictive analytics in combination with Internet of Things (IoT) technologies. This subchapter delves into the potential of predictive analytics for crop yield optimization and highlights its significance in the Farm of the Future.

Predictive analytics, a branch of advanced analytics, involves the use of historical data, statistical algorithms, and machine learning techniques to predict future outcomes. In the context of agriculture, predictive analytics can revolutionize the way farmers make decisions by providing valuable insights into crop growth patterns, pest and disease management, yield forecasting, and resource optimization.

By harnessing the power of IoT technologies, which connect physical devices and sensors to the internet, farmers can collect real-time data on various environmental factors such as temperature, humidity, soil moisture, and sunlight. This data, when combined with historical records, can be analyzed using predictive analytics algorithms to generate accurate predictions and actionable recommendations.

The benefits of using predictive analytics for crop yield optimization are manifold. Firstly, it allows farmers to make informed decisions about when to plant, irrigate, fertilize, and harvest, thereby maximizing crop yield and minimizing resource wastage. Secondly, it enables early detection and prevention of pest infestations and diseases, reducing the need for chemical interventions and promoting sustainable farming practices. Additionally, predictive analytics can aid in predicting market demand, allowing farmers to align their production accordingly and optimize profits.

The potential audience for this subchapter is everyone, ranging from farmers to agricultural researchers, technology enthusiasts, and policymakers. Regardless of their level of knowledge in the field of IoT, the content provides a comprehensive overview of the significance of predictive analytics in optimizing crop yield.

With the rapid advancements in IoT technologies and data analytics, the future of agriculture holds immense promise. Predictive analytics, combined with IoT, offers an unprecedented opportunity to revolutionize the way we approach farming and ensure sustainable food production. By embracing these innovative solutions, we can pave the way for a more efficient, productive, and environmentally conscious Farm of the Future.

# Chapter 4: IoT Applications for Livestock Tracking and Management

**Importance of Livestock Tracking in Agriculture**

Introduction:

Livestock tracking is a crucial aspect of modern agriculture that utilizes Internet of Things (IoT) technologies to optimize livestock management. With the growing demand for food production and the need to ensure animal welfare, tracking livestock has become an essential practice for farmers and ranchers. This subchapter will delve into the significance of livestock tracking in agriculture and how IoT technologies are revolutionizing this field.

Enhancing Animal Welfare:

One of the primary reasons why livestock tracking is vital is its ability to improve animal welfare. By using IoT devices such as GPS trackers, farmers can monitor the location, behavior, and health of their livestock in real-time. This allows them to promptly identify any signs of distress or illness, ensuring timely intervention and reducing the risk of severe health issues. Additionally, tracking livestock can help prevent theft and enable quick recovery of stolen animals, further promoting their well-being.

Optimizing Grazing Patterns:

Efficient pasture management is crucial for maintaining the health and productivity of livestock. With IoT-based tracking systems, farmers can accurately monitor grazing patterns and rotational grazing practices. By analyzing data collected from the GPS trackers, farmers can determine the optimal time to rotate their animals to different grazing areas, preventing overgrazing and soil degradation. This promotes sustainable agriculture practices and improves the quality of the forage available to the livestock.

Improving Breeding and Reproduction:

Tracking individual livestock allows farmers to closely monitor breeding and reproduction cycles. IoT technologies enable the collection of data on estrus behavior, mating patterns, and gestation periods. By analyzing this information, farmers can optimize breeding programs, ensuring healthier offspring and maximizing genetic diversity. This not only improves the overall quality of the livestock but also enhances breeding efficiency, ultimately leading to increased productivity and profitability.

Preventing Disease Outbreaks:

Disease outbreaks can cause significant losses in the livestock industry. However, with livestock tracking, farmers can detect early signs of disease transmission or infection. By continuously monitoring the health parameters of individual animals, such as body temperature and activity levels, farmers can identify anomalies that may indicate the onset of illness. This enables prompt isolation of affected animals, preventing the spread of disease and minimizing economic losses.

Conclusion:

Livestock tracking through IoT technologies has revolutionized the agricultural industry by optimizing animal welfare, grazing patterns, breeding programs, and disease prevention. The ability to monitor and manage livestock in real-time allows farmers to make informed decisions, leading to increased productivity, profitability, and sustainability in agriculture. By embracing IoT-based livestock tracking, farmers can truly harness the power of technology to create a farm of the future.

## IoT-Based Animal Monitoring Systems

In recent years, the Internet of Things (IoT) has revolutionized various industries, and agriculture is no exception. One of the most significant advancements in this field is the development of IoT-based animal monitoring systems. These innovative technologies empower farmers and animal caretakers to keep a closer eye on their livestock, ensuring their well-being and optimizing their productivity.

IoT-based animal monitoring systems utilize sensors and devices to collect data on various aspects of an animal's health and behavior. These sensors can be attached to animals' collars or implanted directly into their bodies, depending on the specific application. The data collected is then transmitted wirelessly to a central database or cloud platform, where it can be analyzed and interpreted.

The benefits of IoT-based animal monitoring systems are numerous. Firstly, they provide real-time information on an animal's vital signs, such as heart rate, body temperature, and respiratory rate. This allows early detection of any health issues or abnormalities, enabling prompt intervention and reducing the risk of diseases spreading among the livestock.

Furthermore, these systems offer insights into an animal's behavior patterns, such as feeding habits, movement, and social interactions. By analyzing this data, farmers can gain a deeper understanding of their animals' well-being and make informed decisions to optimize their care. For instance, they can adjust feeding schedules, identify potential stressors, or detect signs of distress.

In addition to animal welfare, IoT-based animal monitoring systems also contribute to improved productivity and efficiency on farms. By monitoring an animal's health and behavior, farmers can identify optimal breeding times, detect pregnancy, and predict the onset of labor. This enables them to maximize breeding success rates and plan for the arrival of new offspring.

Moreover, these systems can detect changes in an animal's feeding patterns, which may indicate changes in appetite or overall health. By addressing such issues promptly, farmers can ensure their livestock maintains optimal nutrition, leading to better growth rates and higher-quality produce.

Overall, IoT-based animal monitoring systems are a game-changer in the agricultural industry. By integrating IoT technologies into animal management practices, farmers can enhance animal welfare, improve productivity, and optimize their overall operations. These systems empower farmers with valuable insights, enabling them to make data-driven decisions and ensure the well-being and success of their livestock. With the continued advancements in IoT, we can expect even more sophisticated and efficient animal monitoring systems in the farm of the future.

## Smart Feeding and Health Management Solutions

In the ever-evolving world of agriculture, the implementation of Internet of Things (IoT) technologies has revolutionized the way farmers manage their operations. One area where IoT has made a significant impact is in the realm of smart feeding and health management solutions for livestock. By harnessing the power of connected devices and data analytics, farmers can now optimize the feeding process and ensure the overall health and well-being of their animals like never before.

Gone are the days of manual feeding and guesswork. With smart feeding solutions, farmers can now accurately measure and control the amount of feed given to each animal, minimizing waste and maximizing efficiency. IoT-enabled feeders can be programmed to dispense precise portions of feed based on individual animal needs, taking into account factors like weight, age, and health conditions. This not only eliminates overfeeding but also ensures that every animal receives the necessary nutrients for optimal growth and productivity.

Furthermore, IoT technologies have paved the way for real-time monitoring of animal health. By equipping livestock with smart sensors and wearables, farmers can continuously track vital parameters such as body temperature, heart rate, and activity levels. This data is then transmitted to a central system where it is analyzed to detect any signs of illness or distress. Prompt alerts are sent to the farmer, enabling early intervention and preventive measures to be taken. With such proactive health management, farmers can significantly reduce the risk of disease outbreaks and improve the overall welfare of their animals.

The benefits of smart feeding and health management solutions extend beyond individual farms. By aggregating and analyzing data from multiple farms, patterns and trends can be identified, leading to better disease prevention strategies and improved animal breeding programs at a larger scale. Additionally, by optimizing feeding practices, farmers can reduce their environmental footprint by minimizing feed waste and optimizing resource utilization.

In conclusion, IoT technologies have brought about a paradigm shift in the way livestock is fed and managed. Smart feeding and health management solutions have not only increased efficiency and productivity but have also improved animal welfare and reduced environmental impact. As the world continues to face challenges in meeting the growing demand for food, these IoT-enabled solutions are crucial in optimizing agriculture and paving the way for a sustainable and technologically advanced future.

**Improving Livestock Productivity with IoT Technologies**

In recent years, the Internet of Things (IoT) has revolutionized various industries, and agriculture is no exception. With the advent of IoT technologies, farmers and livestock owners have gained access to advanced tools and systems to optimize their operations and enhance productivity. This subchapter explores the significant role IoT technologies play in improving livestock productivity and the overall efficiency of farming practices.

Livestock management is a complex task that requires constant monitoring and control. IoT technologies offer a range of solutions that enable real-time data collection, analysis, and automation, leading to better decision-making and streamlined processes. By leveraging IoT devices such as smart sensors, farmers can effortlessly monitor the health, behavior, and environmental conditions of their livestock.

Through the use of smart collars or ear tags embedded with sensors, farmers can track the location and movement patterns of their livestock, ensuring their safety and preventing theft. These sensors can also monitor vital signs, such as body temperature and heart rate, enabling early detection of illnesses and allowing prompt intervention. This preventive approach not only improves animal welfare but also minimizes the risk of disease outbreaks, ultimately increasing the overall productivity of the herd.

Additionally, IoT technologies enable farmers to optimize feeding regimes, ensuring each animal receives the appropriate nutrition for their needs. Smart feeding systems can be utilized to monitor individual consumption patterns, detect feeding anomalies, and adjust feed distribution accordingly. This level of precision ensures optimal growth rates, reduces feed wastage, and saves costs.

Efficient reproduction management is essential for livestock productivity. IoT technologies facilitate automated monitoring of estrus cycles in animals, enabling accurate breeding timing. By monitoring hormone levels and other reproductive indicators, farmers can maximize the success rate of insemination and improve overall breeding efficiency.

Moreover, IoT technologies enable remote monitoring of livestock facilities, ensuring optimal living conditions. Smart ventilation systems, for instance, can adjust airflow and temperature based on real-time data, preventing heat stress and reducing energy consumption. Automated water management systems can also ensure a constant supply of clean water, minimizing the risk of dehydration and improving animal health.

In conclusion, IoT technologies have revolutionized the agriculture industry, particularly in livestock management. By harnessing the power of real-time data, automation, and smart sensors, farmers can optimize their operations, improve animal welfare, and increase overall productivity. The integration of IoT technologies into livestock management practices not only benefits farmers but also ensures sustainable and efficient food production for a growing global population.

# Chapter 5: Data Analytics and Decision Support Systems for Smart Agriculture

## Collecting and Analyzing Agricultural Data

In today's digital age, the Internet of Things (IoT) has revolutionized various industries, and agriculture is no exception. The ability to collect and analyze agricultural data through IoT technologies has opened up new possibilities for optimizing farming practices and maximizing yields. This subchapter will delve into the significance of collecting and analyzing agricultural data, exploring its benefits and providing insights into how IoT technologies are transforming the agricultural landscape.

Agricultural data collection involves gathering information from various sources such as sensors, drones, satellites, and weather stations. These devices continuously monitor factors like soil moisture, temperature, humidity, crop growth, and pest infestations, providing real-time data that farmers can utilize to make informed decisions. By leveraging IoT technologies, farmers can now collect data from vast areas of farmland and analyze it to gain valuable insights.

One of the key benefits of collecting agricultural data is its potential to enhance crop yield and quality. By monitoring factors such as soil moisture and nutrient levels, farmers can optimize irrigation and fertilization, ensuring crops receive the precise amount of resources they need. This data-driven approach allows for more efficient use of resources, reducing waste and increasing productivity.

Another advantage of agricultural data analysis is its ability to predict and prevent crop diseases and pest infestations. By monitoring environmental conditions and using predictive analytics, farmers can identify potential threats and take proactive measures to protect their crops. This not only saves valuable resources but also ensures a healthy harvest.

Furthermore, by analyzing historical data, farmers can gain insights into long-term trends and patterns. This information is invaluable for making informed decisions about crop rotation, choosing the right varieties, and even predicting market demands. IoT technologies enable farmers to access this data easily and make data-driven decisions that can contribute to the overall success of their farming operations.

In conclusion, the collection and analysis of agricultural data through IoT technologies have transformed the way farmers operate. By leveraging real-time data, farmers can optimize their resources, predict and prevent crop diseases, and make informed decisions. The use of IoT in agriculture has the potential to revolutionize farming practices and pave the way for a more sustainable and efficient future. Whether you are a farmer, an agriculture enthusiast, or simply interested in the Internet of Things, understanding the significance of collecting and analyzing agricultural data is crucial for embracing the future of farming.

## Applying Machine Learning and AI in Agriculture

In recent years, the agricultural industry has witnessed a remarkable transformation with the advent of advanced technologies such as the Internet of Things (IoT), Machine Learning (ML), and Artificial Intelligence (AI). These cutting-edge innovations have paved the way for the development of smart farming techniques, enabling farmers to optimize their yields, enhance sustainability, and streamline overall operations.

Machine Learning and AI have revolutionized the way agriculture is approached, empowering farmers to make data-driven decisions and leverage predictive analytics for improved efficiency. By collecting and analyzing vast amounts of data from various IoT devices, such as weather sensors, soil moisture monitors, and drones, ML algorithms can identify patterns, detect anomalies, and provide valuable insights to farmers.

One of the key applications of ML and AI in agriculture is precision farming. By combining real-time data with historical information, farmers can precisely determine the optimal amount of water, fertilizers, and pesticides required for specific areas of their fields. This targeted approach not only reduces resource wastage but also minimizes environmental impact, making agriculture more sustainable.

Moreover, ML algorithms can assist in crop disease detection and prevention. By analyzing images of plants captured by drones or sensors, AI models can identify early signs of diseases or nutrient deficiencies, enabling farmers to take timely action and prevent the spread of infections. This proactive approach helps farmers save crops and improve overall productivity.

Additionally, ML algorithms can aid in yield prediction, guiding farmers in making informed decisions regarding harvesting and sales. By considering factors such as weather conditions, soil quality, and historical data, AI models can estimate crop yields accurately. This information helps farmers plan their logistics, optimize storage facilities, and negotiate better prices with buyers.

Furthermore, ML algorithms can automate tasks such as irrigation scheduling, weed control, and pest management. By integrating IoT devices and AI-powered systems, farmers can reduce manual labor, save time, and focus on more critical aspects of their operations. This automation not only improves productivity but also enhances the overall efficiency of the agricultural process.

In conclusion, the integration of Machine Learning and AI in agriculture has brought about a paradigm shift in the industry. With the aid of IoT technologies, farmers can now harness the power of data to optimize their farming practices. From precision farming to disease detection and yield prediction, these advancements enable farmers to make smarter decisions, increase productivity, and contribute to sustainable agriculture practices. As we move towards the Farm of the Future, it is crucial for farmers and agriculture enthusiasts to embrace these technologies and leverage their potential for a more efficient and thriving agricultural sector.

## Smart Farming Decision Support Systems

In recent years, the advent of Internet of Things (IoT) technologies has revolutionized various industries, and agriculture is no exception. The integration of IoT into farming practices has given rise to the concept of "Smart Farming," which leverages advanced data analytics and automation to optimize agricultural operations. One crucial aspect of this technological transformation is the use of Decision Support Systems (DSS) in farming.

Smart Farming Decision Support Systems are software applications that enable farmers to make informed decisions by analyzing real-time data collected from sensors and other IoT devices. These systems provide farmers with valuable insights into various aspects of their farming operations, including soil health, weather conditions, crop growth, and livestock management. By harnessing the power of IoT, DSS empowers farmers to make data-driven decisions that can significantly improve productivity, reduce costs, and minimize environmental impact.

One of the key benefits of Smart Farming Decision Support Systems is their ability to provide accurate and timely information to farmers. By gathering data from multiple sources, such as weather stations, soil sensors, and crop monitoring devices, DSS can offer farmers real-time updates on critical factors affecting their farming practices. This information allows farmers to make informed decisions, such as when to irrigate, fertilize, or harvest their crops, ensuring optimal yield and resource utilization.

Moreover, Smart Farming Decision Support Systems enable farmers to adopt precision agriculture techniques. By utilizing advanced analytics and machine learning algorithms, DSS can identify patterns and trends in farming data, allowing farmers to implement precise and site-specific interventions. This not only prevents overuse of resources but also minimizes the use of pesticides and fertilizers, leading to sustainable and environmentally friendly farming practices.

Additionally, DSS can enhance livestock management by monitoring animal health, tracking feed consumption, and predicting disease outbreaks. By continuously monitoring vital signs and behavior patterns, farmers can detect early signs of illness and take prompt action, minimizing the risk of disease spread and ensuring the well-being of their livestock.

In conclusion, Smart Farming Decision Support Systems are a game-changer in the agriculture industry, leveraging IoT technologies to optimize farming practices. By providing farmers with real-time and accurate data, DSS enables them to make informed decisions, implement precision agriculture techniques, and enhance livestock management. As the world faces increasing food demand and environmental challenges, Smart Farming DSS offers a promising solution to ensure sustainable and efficient agriculture for everyone.

**Benefits and Challenges of Data-Driven Agriculture**

In today's rapidly evolving world, the Internet of Things (IoT) has emerged as a game-changer in various industries, and agriculture is no exception. With the integration of IoT technologies, data-driven agriculture has become a promising avenue to optimize farming practices and improve overall productivity. This subchapter aims to explore the benefits and challenges associated with data-driven agriculture, shedding light on its potential implications for the future of farming.

One of the primary advantages of data-driven agriculture is its ability to provide farmers with real-time insights and analytics. By collecting and analyzing data from various sources such as weather stations, soil moisture sensors, and crop growth patterns, farmers can make more informed decisions. They can optimize irrigation schedules, apply fertilizers more efficiently, and detect early signs of diseases or pest infestations. This data-driven approach not only enhances crop yields but also minimizes waste and reduces environmental impact.

Furthermore, data-driven agriculture enables farmers to achieve precision farming. By using IoT devices to gather data on soil conditions, temperature, humidity, and other factors, farmers can tailor their farming practices to specific areas of their fields. This targeted approach allows for optimized resource allocation, reducing costs and maximizing crop quality. Additionally, by monitoring and controlling machinery remotely using IoT-enabled devices, farmers can enhance operational efficiency and streamline their workflows.

However, despite its numerous benefits, data-driven agriculture also presents several challenges. One of the key concerns is data security and privacy. As more devices become interconnected, the potential for cyber-attacks and data breaches increases. Protecting sensitive information such as crop data, financial records, and customer information becomes crucial to maintain trust and integrity within the agricultural ecosystem.

Another challenge lies in the implementation and adoption of data-driven agriculture technologies. Many farmers, especially those in developing regions or with limited resources, may face barriers to access and affordability. Ensuring equitable access to IoT technologies and providing adequate training and support will be crucial to bridge this digital divide.

In conclusion, data-driven agriculture holds immense potential to revolutionize the farming industry. It offers a range of benefits, including improved decision-making, increased productivity, and reduced environmental impact. However, challenges such as data security and accessibility need to be addressed to fully harness the power of IoT technologies in agriculture. By overcoming these obstacles, data-driven agriculture can pave the way for a more sustainable and efficient future for farmers worldwide.

# Chapter 6: IoT Integration and Connectivity in Agriculture

## Wireless Sensor Networks for Smart Agriculture

In recent years, the integration of Internet of Things (IoT) technologies has revolutionized various industries, and agriculture is no exception. The emergence of wireless sensor networks (WSNs) has paved the way for smart agriculture, enabling farmers to optimize their operations and make data-driven decisions to enhance productivity and sustainability. This subchapter explores the potential of WSNs in transforming the agricultural landscape and outlines the numerous benefits they offer to farmers and the environment.

WSNs are composed of small, low-cost sensor nodes that are distributed throughout the farm to collect data on various environmental factors such as temperature, humidity, soil moisture, and light intensity. These nodes communicate wirelessly with a central base station, which collects, analyzes, and processes the data, providing farmers with real-time insights into their crops and livestock. This continuous monitoring allows farmers to detect anomalies and make timely interventions, ensuring optimal conditions for plant growth and animal health.

One of the primary advantages of WSNs in agriculture is their ability to conserve resources and reduce environmental impact. By precisely monitoring soil moisture levels, farmers can optimize irrigation, preventing water wastage and reducing energy consumption. Similarly, real-time weather data from WSNs can help farmers schedule their irrigation and fertilization activities, minimizing the use of chemicals and preventing runoff into water bodies. With the precise application of resources, farmers can achieve higher yields while minimizing the negative impacts on the environment.

Moreover, WSNs enable farmers to implement precision agriculture techniques, tailoring their practices to the specific needs of each crop or livestock. By monitoring the microclimate within the field, farmers can identify areas that require attention, such as pest infestations or nutrient deficiencies, and take targeted actions, reducing the need for broad-spectrum pesticides or excessive fertilization. This not only reduces costs but also promotes sustainable farming practices.

In addition to resource optimization, WSNs also enhance farm management by automating routine tasks and providing valuable data-driven insights. For instance, WSNs can monitor livestock behavior, enabling early detection of illness or distress. Similarly, by analyzing historical data, WSNs can help farmers predict crop yields and optimize planting and harvesting schedules.

In conclusion, wireless sensor networks have emerged as a game-changer in the field of agriculture, enabling farmers to embrace smart farming practices. By monitoring environmental parameters, optimizing resource allocation, and automating routine tasks, WSNs empower farmers to make informed decisions, maximize crop yields, minimize environmental impact, and achieve sustainable agricultural practices. With the integration of IoT technologies, the future of farming looks promising, bringing us closer to a more efficient, productive, and sustainable farm of the future.

## IoT Platforms and Cloud-Based Solutions

In recent years, the Internet of Things (IoT) has revolutionized various industries, and agriculture is no exception. IoT technologies have paved the way for the Farm of the Future, where farmers can optimize their agricultural practices and achieve unprecedented levels of efficiency and productivity. Central to this transformation are IoT platforms and cloud-based solutions.

IoT platforms serve as the backbone of any IoT ecosystem. They provide the necessary infrastructure to connect, manage, and analyze data from a multitude of IoT devices deployed on farms. These platforms enable seamless communication between devices, sensors, and actuators, creating a network of interconnected devices that can monitor and control various farm operations. With the help of IoT platforms, farmers can remotely monitor soil moisture levels, weather conditions, crop growth, and livestock health, among many other parameters critical to successful farming.

Cloud-based solutions complement IoT platforms by offering scalable and flexible storage and computing capabilities. Through the cloud, farmers can securely store and analyze vast amounts of data collected from IoT devices. This data can then be processed using advanced analytics techniques, including machine learning and artificial intelligence, to extract valuable insights. By leveraging cloud-based solutions, farmers can make data-driven decisions, optimize resource allocation, and predict and prevent potential issues, such as crop diseases or equipment failures.

The benefits of IoT platforms and cloud-based solutions in agriculture are manifold. Firstly, they enable real-time monitoring and control of agricultural processes, allowing farmers to take immediate action in response to changing conditions. This helps in reducing waste and maximizing resource utilization, leading to higher yields and profitability. Moreover, IoT platforms and cloud-based solutions facilitate data sharing and collaboration between farmers, researchers, and other stakeholders, fostering innovation and knowledge exchange within the agricultural community.

For the everyday consumer, the impact of IoT platforms and cloud-based solutions in agriculture extends beyond the farm. By optimizing agricultural practices, these technologies contribute to sustainable farming methods, reducing environmental impact and promoting food security. Consumers can also benefit from improved food quality and safety, as IoT platforms can track the entire supply chain, ensuring transparency and traceability from farm to table.

In conclusion, IoT platforms and cloud-based solutions are revolutionizing agriculture by enabling the Farm of the Future. These technologies empower farmers to optimize their practices, increase productivity, and ensure sustainable and efficient food production. With IoT platforms and cloud-based solutions, the potential for innovation in agriculture is limitless, promising a brighter and more prosperous future for everyone.

**Integrating IoT Technologies with Existing Farming Systems**

In this subchapter, we will explore the exciting possibilities that arise when integrating IoT technologies with existing farming systems. As the world faces increasing challenges in food production due to population growth and climate change, the Internet of Things (IoT) offers innovative solutions that can revolutionize agriculture and optimize farming practices.

The IoT refers to the network of physical devices, vehicles, and other objects embedded with sensors, software, and connectivity, enabling them to collect and exchange data. By harnessing the power of IoT technologies, farmers can monitor and control various aspects of their operations remotely, leading to improved efficiency, productivity, and sustainability.

One significant benefit of integrating IoT technologies in farming systems is the ability to collect real-time data from sensors placed throughout the farm. These sensors can measure soil moisture levels, air temperature, humidity, and even the health status of livestock. By accessing this data through a centralized platform, farmers gain valuable insights into their operations, enabling them to make informed decisions and take proactive measures to optimize their processes.

For instance, IoT-enabled irrigation systems can use data from soil moisture sensors to deliver water precisely when and where it is needed, minimizing waste and conserving resources. Livestock monitoring systems can alert farmers to any signs of illness or distress, allowing for immediate intervention and preventing potential losses.

Furthermore, IoT technologies facilitate precision agriculture, where farmers can apply inputs such as fertilizers or pesticides precisely where they are needed, reducing costs and environmental impact. Drones equipped with sensors and cameras can provide high-resolution images of crop health, enabling farmers to detect disease or nutrient deficiencies early on and take appropriate action.

The integration of IoT technologies with existing farming systems also opens up opportunities for automation. Automated machinery can perform tasks such as planting, harvesting, and sorting crops, reducing labor costs and increasing productivity. Smart devices can also control and adjust environmental conditions in greenhouses or livestock facilities, ensuring optimal growing conditions and animal welfare.

In conclusion, integrating IoT technologies with existing farming systems has the potential to revolutionize agriculture as we know it. By harnessing the power of real-time data, precision agriculture, and automation, farmers can optimize their operations, increase productivity, and contribute to a more sustainable and resilient food system. The possibilities are endless, and the Farm of the Future will rely heavily on IoT technologies to meet the challenges of tomorrow.

**Ensuring Data Security and Privacy in IoT Agriculture**

In today's digital age, the Internet of Things (IoT) has revolutionized various industries, including agriculture. IoT technologies have paved the way for smarter farming practices, enabling farmers to optimize their operations and increase productivity. However, with the integration of IoT devices and sensors into agriculture, it is crucial to address the concerns surrounding data security and privacy.

Data security and privacy are paramount in any IoT application, and agriculture is no exception. As farmers gather valuable data from their fields through sensors and devices, it is essential to implement robust security measures to protect this information from unauthorized access or tampering. This subchapter aims to shed light on the importance of data security and privacy in IoT agriculture and provide valuable insights on how to ensure their integrity.

First and foremost, it is vital for farmers to invest in secure IoT devices and sensors. Choosing reputable and trustworthy manufacturers that prioritize security is key. Such devices should employ encryption techniques, strong authentication protocols, and regular firmware updates to safeguard data integrity.

Moreover, farmers must establish secure communication channels for transmitting data. Encryption protocols should be implemented to encrypt data during transmission, preventing interception by unauthorized entities. Employing firewalls and intrusion detection systems can further fortify the communication network against potential cyber threats.

In addition to protecting the data during transmission, it is equally important to secure the storage of data. Farmers should adopt secure cloud-based platforms or on-premises servers with robust access controls and backups to ensure data availability, integrity, and confidentiality.

Furthermore, farmers should establish strict access controls to limit data accessibility to authorized personnel only. Implementing multi-factor authentication, strong passwords, and user privilege management will prevent unauthorized individuals from accessing sensitive information.

To enhance data privacy, farmers should anonymize and aggregate data whenever possible. By removing personally identifiable information, farmers can protect the privacy of individuals while still benefiting from valuable insights derived from the aggregated data.

Educating farmers and their workforce about data security best practices is also crucial. Regular training sessions on identifying and mitigating cyber threats, as well as promoting a culture of cybersecurity awareness, can significantly reduce the risk of data breaches.

In conclusion, while IoT technologies offer immense potential to optimize agriculture, ensuring data security and privacy is of paramount importance. By investing in secure devices, establishing robust communication channels, securing data storage, implementing strict access controls, anonymizing data, and educating the workforce, farmers can effectively protect their valuable data. Embracing these measures will not only safeguard the interests of farmers but also contribute to the overall growth and sustainability of IoT agriculture for the benefit of everyone.

# Chapter 7: Case Studies and Success Stories in IoT-Driven Agriculture

## Smart Agriculture Projects in Developed Countries

In recent years, the Internet of Things (IoT) has revolutionized numerous industries, and agriculture is no exception. Developed countries have embraced this technology to optimize their farming practices, resulting in what is now known as smart agriculture. This subchapter explores some remarkable smart agriculture projects implemented in developed countries, showcasing the potential of IoT technologies in transforming the farming industry.

One notable project is the use of sensor networks in precision farming. These networks consist of various sensors such as soil moisture, temperature, and pH sensors, which gather real-time data from the fields. This data is then analyzed and used to provide farmers with valuable insights about crop health, irrigation needs, and nutrient management. By leveraging IoT technologies, farmers can make more informed decisions, leading to increased yields, reduced costs, and minimized environmental impact.

Another fascinating project is the implementation of autonomous farming machinery. Developed countries have started adopting autonomous tractors and drones equipped with advanced sensors and computer vision capabilities. These machines can perform precise tasks such as planting, spraying, and harvesting without human intervention. By utilizing IoT technologies, farmers can achieve higher efficiency, greater precision, and enhanced safety on their farms.

Furthermore, the integration of data analytics and predictive models has proven to be a game-changer in smart agriculture. By analyzing vast amounts of data collected from IoT devices, farmers can gain insights into weather patterns, pest infestations, and disease outbreaks. This information enables them to take proactive measures, such as adjusting planting schedules, applying targeted treatments, and implementing preventive measures, ultimately improving crop yields and reducing losses.

In developed countries, smart agriculture projects have also focused on resource management and sustainability. IoT technologies have been instrumental in optimizing water usage through smart irrigation systems. These systems monitor soil moisture levels and weather conditions in real-time, allowing farmers to precisely control the amount and timing of irrigation. This not only reduces water waste but also ensures that crops receive optimal hydration, resulting in improved water efficiency and conservation.

In conclusion, smart agriculture projects in developed countries have utilized IoT technologies to enhance farming practices in various ways. From precision farming and autonomous machinery to data analytics and resource management, the integration of IoT has significantly transformed the agriculture industry. These advancements not only increase productivity and profitability but also contribute to sustainable and environmentally friendly farming practices. As IoT continues to evolve, the potential for further innovation in smart agriculture is vast, promising a bright future for farmers worldwide.

## IoT Adoption in Developing Nations for Agricultural Growth

In recent years, the Internet of Things (IoT) has revolutionized various industries, and one sector that has greatly benefited from this technology is agriculture. While developed nations have been quick to embrace IoT in their farming practices, developing nations have also recognized its potential and are increasingly adopting these technologies to optimize agricultural growth.

One of the primary challenges faced by developing nations in the agricultural sector is the lack of access to modern farming techniques and resources. Limited infrastructure, unreliable weather conditions, and insufficient knowledge often hinder the productivity and profitability of farmers. However, with the advent of IoT technologies, these challenges are being addressed effectively.

IoT offers a range of solutions that empower farmers in developing nations to enhance agricultural practices and achieve sustainable growth. For instance, smart sensors and weather monitoring devices enable farmers to collect real-time data on temperature, humidity, soil moisture, and rainfall. This data is then analyzed to provide valuable insights, helping farmers make informed decisions regarding irrigation, crop selection, and pest control.

In addition to weather monitoring, IoT-powered precision agriculture systems enable farmers to optimize the use of fertilizers, pesticides, and water. By using connected devices, farmers can precisely target the application of these resources based on the specific needs of their crops, thus reducing waste and minimizing environmental impact. This not only improves productivity but also contributes to sustainable farming practices.

Moreover, IoT technologies empower farmers by providing them with access to relevant information and expert advice. Through connected devices, farmers in remote areas can receive real-time guidance on crop management, disease control, and market prices. This helps them make informed decisions that maximize their yield and profitability.

The adoption of IoT in developing nations for agricultural growth has several benefits beyond individual farmers. It has the potential to improve food security, as increased productivity ensures a consistent food supply. Additionally, it creates new job opportunities and boosts local economies by attracting investments in the agricultural sector.

In conclusion, IoT adoption in developing nations for agricultural growth is a game-changer. By leveraging IoT technologies, farmers can overcome the challenges they face and achieve sustainable growth. The impact of these technologies goes beyond individual farms, benefiting entire communities and contributing to the overall development of these nations. As IoT continues to evolve, it is vital for developing nations to embrace these technologies and unlock their potential for a brighter future in agriculture.

**Lessons Learned and Best Practices from Successful Implementations**

In the rapidly evolving world of agriculture, the integration of Internet of Things (IoT) technologies has revolutionized the way farmers operate and optimize their operations. With the aim of sharing invaluable insights and experiences from successful implementations, this subchapter delves into the lessons learned and best practices that have emerged from the Farm of the Future.

1.          Start          Small,          Scale          Fast: One crucial lesson learned is the importance of starting with a small-scale implementation of IoT technologies. This allows farmers to test the waters, understand potential challenges, and fine-tune their strategies before scaling up. By starting small and gradually expanding, farmers can mitigate risks and optimize the implementation process.

2.          Collaborate          and          Share          Data: Successful implementations have emphasized the significance of collaboration and data sharing. By partnering with experts, farmers can tap into a wealth of knowledge and gain valuable insights into the effective utilization of IoT technologies. Additionally, sharing data with other farmers, researchers, and industry stakeholders fosters collective learning and innovation.

3.          Prioritize          Data          Security: With the integration of IoT technologies, a vast amount of data is generated and collected. To ensure the success of IoT implementations, farmers must prioritize data security. Implementing robust security measures, such as encryption and access controls, safeguards sensitive information and instills confidence among stakeholders.

4.    Embrace    Automation    and    Predictive    Analytics: Among the best practices that have emerged is the need to embrace automation and utilize predictive analytics. IoT technologies enable real-time monitoring and automation of various processes, reducing manual labor and increasing operational efficiency. Furthermore, predictive analytics allows farmers to gain valuable insights, enabling them to make data-driven decisions and optimize their operations.

5.    Continuous    Monitoring    and    Maintenance: Successful implementations have highlighted the importance of continuous monitoring and maintenance of IoT systems. Regularly assessing the performance of sensors, networks, and software ensures that any issues are promptly addressed, minimizing disruptions and maximizing the benefits derived from IoT technologies.

6.    Invest    in    Training    and    Education: To fully leverage IoT technologies, farmers must invest in training and education for themselves and their workforce. This empowers farmers to navigate the complexities of IoT implementations and utilize these technologies to their fullest potential.

In conclusion, the lessons learned and best practices from successful implementations of IoT technologies in agriculture are invaluable resources for farmers and industry stakeholders alike. By starting small, collaborating, prioritizing data security, embracing automation and predictive analytics, continuously monitoring and maintaining systems, and investing in training and education, farmers can optimize their operations and pave the way for a more sustainable and efficient Farm of the Future.

# Chapter 8: Challenges and Future Directions in IoT-Driven Agriculture

**Addressing Connectivity Issues in Rural Areas**

In today's digital age, connectivity plays a crucial role in various aspects of our lives, including agriculture. However, one significant challenge faced by farmers in rural areas is the lack of reliable internet connectivity. This subchapter aims to address this issue and explore possible solutions to bridge the connectivity gap in rural regions, ultimately optimizing agriculture with IoT technologies.

Rural areas often suffer from limited or no access to high-speed internet, hindering farmers from taking advantage of the latest IoT innovations. This lack of connectivity can have adverse effects on farm productivity, efficiency, and overall profitability. For instance, farmers may struggle to monitor crop health in real-time, access weather forecasts, or collect data on soil conditions, hampering their decision-making process.

To address these challenges, it is essential to understand the underlying causes of connectivity issues in rural areas. One primary reason is the geographical remoteness of these regions, making it economically unfeasible for internet service providers to invest in infrastructure development. Additionally, the topography of rural areas can pose obstacles to signal propagation, further exacerbating the problem.

Fortunately, several solutions can help overcome these connectivity issues. One approach is the use of satellite internet technology, which can provide reliable connectivity regardless of geographical location.

This technology utilizes satellites orbiting the Earth to transmit and receive data, ensuring that even the most remote farms can access the internet. Although satellite internet can be expensive, advancements in technology have made it more affordable and accessible in recent years.

Another solution is the deployment of low-power wide-area networks (LPWANs), specifically designed to cater to IoT applications. LPWANs offer long-range connectivity with low power consumption, making them ideal for rural areas. By utilizing LPWAN technology, farmers can connect their IoT devices, such as sensors and drones, to monitor and manage their farms efficiently.

Furthermore, government initiatives and partnerships between private entities and local communities can play a significant role in addressing connectivity issues. By investing in infrastructure development and collaborating with internet service providers, governments can ensure that connectivity becomes a reality for farmers in rural areas.

In conclusion, addressing connectivity issues in rural areas is vital for optimizing agriculture with IoT technologies. By exploring solutions such as satellite internet and LPWANs, and fostering collaborations between stakeholders, we can bridge the connectivity gap and empower farmers with the tools they need to thrive in the digital era. It is crucial that we prioritize connectivity in rural regions to unlock the full potential of IoT in agriculture and ensure sustainable farming practices for the future.

**Overcoming Adoption Barriers for Small-Scale Farmers**

In today's rapidly advancing world, the Internet of Things (IoT) technologies have the potential to revolutionize the agricultural industry. However, the adoption of these technologies by small-scale farmers is often hindered by various barriers. This subchapter aims to shed light on these challenges and provide insights on how to overcome them, ensuring that even small-scale farmers can benefit from IoT innovations.

One of the primary barriers faced by small-scale farmers is the lack of awareness and understanding of IoT technologies. Many farmers are not aware of the potential benefits that IoT can offer, such as increased productivity, resource optimization, and improved decision-making. Therefore, educating farmers about the advantages and practical applications of IoT technologies is crucial. This can be achieved through workshops, training programs, and information campaigns targeted specifically at small-scale farmers.

Another significant barrier is the cost associated with implementing IoT solutions. Small-scale farmers often operate on a tight budget and may find it challenging to invest in expensive IoT devices and infrastructure. To overcome this barrier, it is essential to develop affordable IoT solutions specifically designed for small-scale farming. Governments, NGOs, and private sector organizations can play a crucial role in providing financial support, subsidies, or incentives to make these technologies more accessible to small-scale farmers.

Limited access to reliable internet connectivity is also a barrier that needs to be addressed. IoT technologies heavily rely on internet connectivity to collect and transmit data. However, many rural areas, where small-scale farmers are predominantly located, suffer from poor or no internet connectivity. To overcome this challenge, alternative connectivity solutions such as low-power, wide-area networks (LPWANs) or satellite-based internet can be explored. Additionally, collaborations with telecommunication companies and government initiatives to improve rural internet infrastructure can help bridge the connectivity gap.

Lastly, the complexity of IoT technologies can deter small-scale farmers from adopting them. Many farmers may lack the technical skills and knowledge required to set up and operate IoT devices effectively. To address this, user-friendly IoT solutions with intuitive interfaces should be developed. Furthermore, providing ongoing technical support, training, and guidance can empower small-scale farmers to overcome any technical challenges they may encounter.

By addressing these adoption barriers, small-scale farmers can fully embrace IoT technologies and reap the benefits they offer. The widespread adoption of IoT in agriculture has the potential to transform small-scale farming, improving productivity, sustainability, and ultimately, the livelihoods of farmers worldwide. It is essential that all stakeholders come together to create an enabling environment that empowers small-scale farmers to embrace the Farm of the Future powered by IoT technologies.

## Ethical and Sustainable Use of IoT Technologies in Agriculture

In recent years, the Internet of Things (IoT) has revolutionized various industries, and agriculture is no exception. With the advent of IoT technologies, farmers and agricultural professionals have gained access to a myriad of tools and data that can optimize their operations and increase productivity. However, as with any technological advancement, the ethical and sustainable use of IoT technologies in agriculture must be carefully considered.

One crucial aspect of the ethical use of IoT technologies in agriculture is data privacy and security. IoT devices collect vast amounts of data, ranging from soil moisture levels to livestock health information. Farmers must ensure that this data is protected from unauthorized access and use. It is essential to implement robust cybersecurity measures to safeguard this sensitive information. Additionally, farmers should be transparent with their customers and stakeholders about the data collected and how it is used, ensuring that privacy concerns are addressed.

Another ethical consideration is the potential impact of IoT technologies on labor practices. While IoT devices can automate certain tasks, such as irrigation or pest control, it is crucial to ensure that this does not lead to job displacement or exploitation. Farmers should strive to provide fair employment opportunities and support the development of skills needed to work with IoT technologies. Additionally, governments and agricultural organizations should collaborate to establish guidelines that prevent the misuse of IoT devices and protect workers' rights.

Sustainability is another vital aspect of using IoT technologies in agriculture. By leveraging IoT devices, farmers can monitor resource usage more efficiently, reduce waste, and enhance overall sustainability. For example, IoT-enabled sensors can monitor soil moisture, enabling farmers to optimize irrigation practices and conserve water. Similarly, IoT devices can monitor energy usage, helping farmers reduce their carbon footprint. Sustainable farming practices are essential for preserving the environment and ensuring the long-term viability of the agricultural industry.

To promote the ethical and sustainable use of IoT technologies in agriculture, it is crucial for farmers, policymakers, and technology providers to collaborate. Farmers should be educated about the potential benefits and risks of IoT technologies, enabling them to make informed decisions. Policymakers should develop regulations that promote responsible IoT use and protect the rights of farmers and consumers. Technology providers should prioritize incorporating ethical design principles and sustainability features into their IoT devices.

Ultimately, the ethical and sustainable use of IoT technologies in agriculture holds immense potential for optimizing farming practices and addressing global food security challenges. By prioritizing data privacy, fair labor practices, and sustainability, the agricultural industry can harness the power of IoT technologies while ensuring a better future for farming and the environment.

**Future Trends and Innovations in Smart Agriculture**

Introduction:
The world of agriculture is entering a new era with the emergence of Internet of Things (IoT) technologies. These technologies are revolutionizing farming practices, enabling farmers to optimize their operations, increase productivity, and make informed decisions. In this subchapter, we will explore the future trends and innovations in smart agriculture, showcasing the potential of IoT to transform farming for the better.

1.	Precision	Farming:
Precision farming is a revolutionary concept that utilizes IoT technologies to monitor and manage crops on a micro-level. By deploying sensors, drones, and satellite imagery, farmers can collect real-time data on soil moisture, temperature, and nutrient levels. This data is then analyzed to provide precise recommendations for irrigation, fertilization, and pest control. Precision farming not only saves resources but also enhances crop yields and promotes sustainable farming practices.

2.	Automated	Farming:
Automation is the way forward for the agriculture industry. IoT-enabled devices and robots are being deployed to perform tasks such as seeding, planting, harvesting, and monitoring crop growth. These machines are equipped with sensors and cameras that capture data about crop health, growth rates, and pest infestations. Farmers can remotely control and monitor these machines, reducing labor costs and optimizing productivity.

3.	Smart	Irrigation:
Water scarcity is a pressing issue in many agricultural regions. Smart

irrigation systems leverage IoT technologies to ensure optimal water usage. Sensors monitor soil moisture levels, weather conditions, and crop water requirements, enabling farmers to irrigate their fields precisely. By avoiding over- or under-irrigation, farmers can conserve water resources, reduce costs, and prevent crop diseases caused by water stress.

4.                          Livestock                          Monitoring:
IoT technologies are also transforming livestock management. Sensors attached to animals can monitor their health, temperature, and location in real-time. This data helps farmers detect diseases, track animal behavior, and ensure the overall well-being of their livestock. By preventing diseases and optimizing breeding practices, farmers can improve animal productivity and welfare.

Conclusion:
The future of agriculture lies in the integration of IoT technologies. Precision farming, automation, smart irrigation, and livestock monitoring are just a few of the innovations that will revolutionize the agriculture industry. By adopting these technologies, farmers can increase productivity, reduce costs, and contribute to sustainable farming practices. It is essential for every farmer and industry stakeholder to embrace the potential of IoT in agriculture to ensure a prosperous and sustainable future.

# Chapter 9: Conclusion and Call to Action

## Recap of Key Points

In this subchapter, we will recap the key points discussed throughout the book "Farm of the Future: Optimizing Agriculture with IoT Technologies." Whether you are an expert in the Internet of Things (IoT) or a newcomer to the field, this recap will serve as a comprehensive summary of the major concepts covered.

First and foremost, we explored the fundamental concept of IoT and its relevance in agriculture. IoT refers to the network of interconnected devices and sensors that collect and exchange data. In the context of farming, IoT technologies play a crucial role in optimizing agricultural practices, improving efficiency, and ensuring sustainable food production.

We then delved into the various applications of IoT in agriculture. From precision farming and smart irrigation to livestock monitoring and crop management, IoT offers a wide range of solutions that enable farmers to make data-driven decisions, minimize resource wastage, and enhance productivity. These applications are not only beneficial for large-scale commercial farms but also for small-scale and urban farmers.

Next, we discussed the key components of an IoT-enabled farm. This includes sensors, actuators, connectivity solutions, and cloud-based platforms for data storage and analysis. The integration of these components allows farmers to monitor environmental conditions, track crop health, automate tasks, and remotely control farm operations.

Furthermore, we emphasized the significance of data analytics and predictive modeling in IoT-enabled agriculture. The data collected from sensors and devices can be analyzed to gain valuable insights, predict crop yields, detect diseases or pests, and optimize resource allocation. Leveraging these insights, farmers can make informed decisions that lead to higher yields, reduced costs, and improved sustainability.

Throughout the book, we also highlighted the potential challenges and risks associated with implementing IoT technologies in agriculture. These include data security and privacy concerns, the need for reliable connectivity in rural areas, and the initial investment required to adopt IoT solutions. However, we stressed that the benefits far outweigh the challenges, and with careful planning and implementation, IoT can revolutionize the agricultural industry.

In conclusion, "Farm of the Future: Optimizing Agriculture with IoT Technologies" has provided a comprehensive overview of how IoT can be utilized to transform farming practices. The key takeaway is that IoT offers unprecedented opportunities for farmers to optimize their operations, increase productivity, and contribute to sustainable food production. By embracing IoT technologies, agriculture can evolve and thrive in the digital era, benefitting not only farmers but also the environment and society as a whole.

# The Role of Individuals and Organizations in the Future of Agriculture

In today's rapidly advancing digital age, the Internet of Things (IoT) has emerged as a game-changer across various industries, and agriculture is no exception. With the increasing global population and the need for sustainable food production, the future of agriculture heavily relies on the active participation of individuals and organizations. This subchapter explores the crucial role that these stakeholders play in optimizing agriculture through IoT technologies.

Individuals, as consumers, have the power to drive change in the agricultural industry. By making informed choices about the food they consume, individuals can influence the demand for sustainable and IoT-driven agriculture. Supporting local farmers who adopt IoT technologies and sustainable practices, such as precision farming and smart irrigation systems, can go a long way in promoting a more efficient and environmentally friendly future of agriculture. Additionally, individuals can actively engage in educating themselves and others about the benefits of IoT technologies in agriculture, helping to create awareness and drive adoption.

On the other hand, organizations, both in the private and public sectors, have a crucial role to play in shaping the future of agriculture. Private companies can invest in research and development to create innovative IoT solutions tailored to the specific needs of farmers. By collaborating with farmers, these organizations can ensure that their solutions are practical, cost-effective, and aligned with the needs of the agricultural community. Partnerships between technology companies and agricultural organizations can also facilitate the integration of IoT technologies into existing farming practices, promoting widespread adoption.

Government bodies and policy-makers have an equally vital role in fostering the future of agriculture. By creating supportive policies and regulations, governments can incentivize and facilitate the adoption of IoT technologies in farming. This includes providing funding opportunities, promoting research and development, and establishing standards to ensure interoperability and data security. Moreover, governments can play a key role in bridging the digital divide by investing in rural connectivity, ensuring that all farmers can access and benefit from IoT technologies.

In conclusion, the future of agriculture relies on the active involvement of individuals and organizations. By making informed choices, supporting local farmers, and advocating for sustainable practices, individuals can drive the demand for IoT-driven agriculture. Meanwhile, organizations, both private and public, can invest in research and development, collaborate with farmers, and shape supportive policies to accelerate the adoption of IoT technologies. Together, these stakeholders can transform agriculture into a more efficient, sustainable, and technology-driven industry, ensuring food security and a better future for all.

## Embracing IoT Technologies for Sustainable and Efficient Farming

In recent years, the Internet of Things (IoT) has revolutionized various industries, and agriculture is no exception. The integration of IoT technologies in farming has paved the way for sustainable and efficient practices that hold great potential for the future of agriculture. This subchapter will explore the ways in which IoT technologies can optimize farming techniques and address the challenges faced by the agricultural industry.

One of the key benefits of IoT technologies in farming is the ability to collect and analyze real-time data. Through sensors and connected devices, farmers can monitor various aspects of their operations, such as soil moisture levels, weather conditions, and livestock health. This data-driven approach allows for precise decision-making, enabling farmers to optimize resource allocation, reduce waste, and increase productivity. By leveraging IoT technologies, farmers can ensure that their farming practices are sustainable, minimizing the use of water, pesticides, and fertilizers while maximizing crop yields.

IoT technologies also play a crucial role in improving the overall efficiency of farming operations. Automated systems can be implemented to control irrigation, fertilization, and pest management, reducing the need for manual labor and minimizing human error. With the help of IoT-enabled devices, farmers can remotely control and monitor their farming operations, enhancing productivity and freeing up valuable time for other essential tasks.

Furthermore, IoT technologies facilitate the integration of precision agriculture methods. By combining data from multiple sources, farmers can create detailed maps of their fields, identifying areas that require specific attention. This targeted approach enables farmers to apply resources precisely where they are needed, reducing costs and minimizing environmental impact. Additionally, the use of drones and satellite imagery, connected through IoT, allows for efficient crop monitoring, detecting diseases and pests at an early stage for prompt action.

The adoption of IoT technologies in agriculture is not limited to large-scale farms; it can be implemented by small-scale farmers as well. With the increasing availability and affordability of IoT devices, farmers from all backgrounds can benefit from the advantages they offer. This democratization of technology empowers every farmer to optimize their practices and contribute to sustainable agriculture.

In conclusion, embracing IoT technologies in farming holds immense potential for sustainable and efficient agriculture. By leveraging real-time data, automation, and precision agriculture methods, farmers can optimize resource allocation, reduce waste, increase productivity, and minimize environmental impact. The integration of IoT technologies in farming is not only beneficial for individual farmers but also for the agricultural industry as a whole. Embracing IoT technologies is a crucial step towards creating a farm of the future that prioritizes sustainability and efficiency for the betterment of the entire world.